Weekly Lesson Plan

This Weekly Lesson Plan Belongs To :

....

Weekly Lesson Plan

Name

Date

Monday	Tuesday	Wednesday

Thursday	Friday	Saturday

Notes

Weekly Lesson Plan

Name	Date

Monday	Tuesday	Wednesday

Thursday	Friday	Saturday

Notes

Weekly Lesson Plan

Name

Date

Monday	Tuesday	Wednesday

Thursday	Friday	Saturday

Notes

Weekly Lesson Plan

Name	Date

Monday	Tuesday	Wednesday

Thursday	Friday	Saturday

Notes

Weekly Lesson Plan

Name	Date

Monday	Tuesday	Wednesday

Thursday	Friday	Saturday

Notes

Weekly Lesson Plan

Name

Date

Monday

Tuesday

Wednesday

Thursday

Friday

Saturday

Notes

Weekly Lesson Plan

Name	Date

Monday	Tuesday	Wednesday

Thursday	Friday	Saturday

Notes

Weekly Lesson Plan

Name

Date

Monday

Tuesday

Wednesday

Thursday

Friday

Saturday

Notes

Weekly Lesson Plan

Name	Date

Monday	Tuesday	Wednesday
Thursday	Friday	Saturday

Notes

Weekly Lesson Plan

Name

Date

Monday	Tuesday	Wednesday

Thursday	Friday	Saturday

Notes

Weekly Lesson Plan

Name

Date

Monday

Tuesday

Wednesday

Thursday

Friday

Saturday

Notes

Weekly Lesson Plan

Name

Date

Monday	Tuesday	Wednesday

Thursday	Friday	Saturday

Notes

Weekly Lesson Plan

Name

Date

Monday	Tuesday	Wednesday

Thursday	Friday	Saturday

Notes

Weekly Lesson Plan

Name	Date

Monday	Tuesday	Wednesday

Thursday	Friday	Saturday

Notes

Weekly Lesson Plan

Name

Date

Monday

Tuesday

Wednesday

Thursday

Friday

Saturday

Notes

Weekly Lesson Plan

Name	Date

Monday	Tuesday	Wednesday

Thursday	Friday	Saturday

Notes

Weekly Lesson Plan

Name

Date

Monday

Tuesday

Wednesday

Thursday

Friday

Saturday

Notes

Weekly Lesson Plan

Name

Date

Monday	Tuesday	Wednesday
Thursday	Friday	Saturday

Notes

Weekly Lesson Plan

Name	Date

Monday	Tuesday	Wednesday

Thursday	Friday	Saturday

Notes

Weekly Lesson Plan

Name	Date

Monday	Tuesday	Wednesday

Thursday	Friday	Saturday

Notes

Weekly Lesson Plan

Name

Date

Monday	Tuesday	Wednesday

Thursday	Friday	Saturday

Notes

Weekly Lesson Plan

Name	Date

Monday	Tuesday	Wednesday

Thursday	Friday	Saturday

Notes

Weekly Lesson Plan

Name

Date

Monday

Tuesday

Wednesday

Thursday

Friday

Saturday

Notes

Weekly Lesson Plan

Name	Date

Monday	Tuesday	Wednesday

Thursday	Friday	Saturday

Notes

Weekly Lesson Plan

Name

Date

Monday	Tuesday	Wednesday

Thursday	Friday	Saturday

Notes

Weekly Lesson Plan

Name

Date

Monday

Tuesday

Wednesday

Thursday

Friday

Saturday

Notes

Weekly Lesson Plan

Name

Date

Monday	Tuesday	Wednesday

Thursday	Friday	Saturday

Notes

Weekly Lesson Plan

Name	Date

Monday	Tuesday	Wednesday

Thursday	Friday	Saturday

Notes

Weekly Lesson Plan

Name	Date

Monday	Tuesday	Wednesday

Thursday	Friday	Saturday

Notes

Weekly Lesson Plan

Name

Date

Monday

Tuesday

Wednesday

Thursday

Friday

Saturday

Notes

Weekly Lesson Plan

Name	Date

Monday	Tuesday	Wednesday

Thursday	Friday	Saturday

Notes

Weekly Lesson Plan

Name	Date

Monday	Tuesday	Wednesday
Thursday	Friday	Saturday

Notes

Weekly Lesson Plan

Name

Date

Monday	Tuesday	Wednesday

Thursday	Friday	Saturday

Notes

Weekly Lesson Plan

Name

Date

Monday	Tuesday	Wednesday

Thursday	Friday	Saturday

Notes

Weekly Lesson Plan

Name	Date

Monday	Tuesday	Wednesday

Thursday	Friday	Saturday

Notes

Weekly Lesson Plan

Name

Date

Monday

Tuesday

Wednesday

Thursday

Friday

Saturday

Notes